PROFITABLE TURKEYS

YOUR GUIDE TO REARING TURKEYS FOR A PROFIT

KATIE THEAR & NIGEL BOWERBANK

www.broadleyspublications.co.uk

PROFIT FROM TURKEYS
YOUR GUIDE TO REARING TURKEYS FOR A PROFIT

Published by Broad Leys Publications Limited

Copyright © 2017 Katie Thear & Nigel Bowerbank
Copyright © 2017 Broad Leys Publications Limited

All rights reserved; no part of this publication may be reproduced, stored in a retrieval system, or transmitted in any form or by any means, electronic, mechanical, photocopying, recording, or otherwise without prior written permission of the publishers.

Although every precaution has been taken to verify the accuracy of the information contained herein, the author and publisher assume no responsibility for any errors or omissions. No liability is assumed for damages that may result from the use of information contained within. The reader should make his/her own evaluation as to the appropriateness or otherwise of any technique described.

A catalogue record for this book is available from the British Library.

ISBN: 978-1-910632-35-2

Third Edition

Front cover photograph:
Bronze Stag Turkey - www.shutterstock.com/Katie Thear

Back cover photographs:
Farm life of white Turkeys - www.shutterstock.com/Kirakos,
A flock of farm turkeys - www.shutterstock.com/Richard Mann,
Domestic turkey, beaing reared on a farm - www.shutterstock.com/erni,
Large roasted turkey free range - www.shutterstock.com/farbled

Other photographs acknowledged on page 39. Unless indicated or credited otherwise, the photographs and illustrations in the book are the author's.

Graphic Design by Roger Bowerbank

Permissions & Advertising
For information about reproducing articles or content from this publication, or to explore the possibility of advertising in or contributing to future editions and other titles by Broad Leys Publications, please visit www.broadleyspublications.co.uk/content

BROAD LEYS
www.broadleyspublications.co.uk

Contents

Foreword	5
Introduction	6
History	7
Regulations	8
Housing	11
Acquiring Stock	15
Feeding	17
Killing and Plucking	20
Selling Christmas Turkeys	23
Profit Sheet	25
Health	28
Breeds	29
Breeding	32
Vermin and Predators	35
Reference Section	36

Foreword

Since the publication of the first and second issues of this book by Katie Thear there has been a surprising resurgence of interest on the part of poultry keepers and the public in less intensively produced Christmas turkeys. It is now no longer an unusual occurrence to find small farmers and smallholders selling 'proper' turkeys.

As well as being updated by Nigel Bowerbank with the latest information, this new edition has a profit sheet, colour photographs and a comprehensive list of suppliers of turkey poults, breeding stock and hatching egg suppliers on a local and national basis with phone numbers and internet contact information where available.

We place great importance on ensuring that our publications are regularly updated, if you feel we have left any important information out of this edition please write and tell us so that it can be included in the future.

Broad Leys Publications Limited

Introduction

Turkey Production is a huge, multi-million, intensive industry where competition is great and only the most efficient and cost-effective operation can survive market forces. Competition from France under-cutting prices of UK turkeys has had a devastating effect on the British turkey industry. So, how can the small producer hope to break even, let alone make a profit in such a competitive situation?

The answer is that he is not in competition with the large producers, for that would be impossible. He is catering for a small, specialized and growing market for freshly killed turkeys which have been reared on a non-intensive basis. The current interest in health and wholefoods has made the public increasingly aware, and uneasy, about some of the practices associated with intensive livestock production. The use of steroids and antibiotics, and the intensive housing of the stock itself, have been the subject of scrutiny by both Royal Commissions and veterinary bodies. Media coverage has also helped to focus attention on the situation.

Whatever the rights and wrongs of the situation, the fact is that the small, home producer is now in a good position to cater for this specialized market, as long as he has a spare building which can be adapted. This is not to say that a complete living is possible from such an enterprise; it should be emphasized that what is referred to here is a part-time operation bringing in a subsidiary income.

History

Turkeys originally came from Mexico and have been in existence for some ten million years, this has been proven by the discovery of fossils from this time.

The scientific name for turkey is Meleagris gallopava, this is Latin gallus and means cock and pavo, or otherwise chicken. However Meleagris was the Roman word for guinea fowl, this shows that the Latin name was confused between chicken and a guinea fowl.

The wild turkey which was bronze in colour had to be able to fly to avoid predators and once flying can glide up to a mile. The early Native American Indians, around 10000AD were able to hunt turkeys for their succulent meat and used the feathers from turkeys as the flight feathers for their arrows. The rest of the bird could be used to make ceremonial clothes, with the spurs from stag birds being used to make arrow heads.

The first six turkeys recorded to enter England were introduced by Mr William Strickland a Yorkshireman in 1526. The birds were purchased or traded for from American Native Indians whilst on his travels in America. A male turkey is called a stag, a female is called a hen and baby turkeys under five to six weeks old are called poults. If many turkeys are moving or grouped together these are called a flock of turkeys. The sounds made by the two sexes are very apparent. The sounds made from the stags are the typical "Gobble, Gobble", whilst the "Click, Click" is the distinctive sound from the hens, the hens make this sound to call one another, or call the baby poults to their parent.

Today some 10 million turkeys are sold per year in the UK for consumption at Christmas, whilst in America, which has a much larger population some 60 million birds are consumed during the Thanksgiving holiday.

We owe a lot to the importers and breeders of the turkey for the many different breeds available.

Regulations

Large producers of turkeys are subject to a wide range of regulations, particularly where frozen turkeys are produced. Small producers are exempt as long as the following conditions are fulfilled:

The birds are raised on their own farms and not bought in from elsewhere. They are sold direct to the consumer, in local markets, or through a local butcher or shop.

This means that freshly killed, gutted and dressed turkeys can be sold at the farm gate, through shops, hotels and restaurants, in the area bounded by one's local authority and the one bordering it in any direction. If you live in an area where the sea provides the boundary on one side, it is worth contacting the Department for Environment Food & Rural Affairs (DEFRA), and asking them to make allowances for this, and to let you sell your dressed birds in two local authority areas bordering you on one side; there is a precedent for this. (The address and telephone number for DEFRA is available on the www.gov.uk website.)

It is also possible to supply retailers further afield with 'Norfolk dressed' turkeys. These are birds which are killed and plucked, but not gutted and

which retain the head and feet. Each bird thus provided for butchers must have an attached label with the name and address of the supplier. The EEC commissioners are not happy about our ability to sell such birds for they say that the dangers of salmonella transference are increased. Salmonella can be found in all poultry, including frozen birds, but thorough cooking destroys it. It is my personal belief that the danger is greater with frozen birds, because of the possibility of inadequate defrosting before cooking, leading to incomplete cooking of the bird. As recently as August 1981, the authorities in Britain decided that the sale of such birds should continue.

The Trade Description Act will apply, so it is obvious that turkeys which are described as freshly killed and organically reared must be so.

Flocks of 50 or more birds must be registered. They don't have to all be the same species. You should register your poultry within one month of their arrival at your premises.

You do not have to register for fewer than 50 birds. However, you are encouraged to register voluntarily so that DEFRA can contact you quickly if there is an outbreak of disease.

The Protection of Animals Act and the Agriculture (Miscellaneous Provisions) Act state that it is an offence to cause unnecessary pain or distress to livestock on agricultural land. Birds which are to be slaughtered are covered by the Slaughter of Poultry Act and Humane Conditions Regulations. These require that poultry awaiting slaughter should not be subjected to unnecessary pain or distress, must be slaughtered as soon as practicable and, meanwhile must be protected from bad weather. Slaughtering must be instantaneous and either be by decapitation or neck dislocation.

General regulations which apply to food hygiene apply, and every effort should be made to protect the fresh carcasses during plucking, and to store them in a cool, protected environment before sale.

Housing

For the small producer, the traditional American pole barn system of housing is the best. This is, essentially, a simple barn structure which allows adequate ventilation and at the same time, protection from the weather. The diagram on page 15 Shows a home-made building, boarded off on two sides to give protection against prevailing winds. A barn can be adapted as long as it is secure to prevent a fox gaining access. The wire mesh on the other two sides allows plenty of fresh air as well as a certain amount of sunlight. Obviously, any substantial outbuilding is suitable for turkeys, as long as there is a good flow of air available. Turkeys are hardy birds and are more likely to develop problems if they are housed in a close, stuffy environment.

Wood shavings or chopped straw provide a suitable litter for them and, as long as the depth is 3"inches (7.6cms), will usually last until the birds are ready for slaughter. If it does become wet or dirty at any stage, it can be added to as necessary, and the whole lot removed for composting when the batch of turkeys are ready to go.

Perches are not necessary but if, like me, you prefer to give the birds as natural environment as possible, the provision of low ones will be readily accepted by them. They must not be high otherwise the bulky birds may find difficulty in getting up. There must be enough room on perches for all the birds, otherwise the birds will not be happy and jostle for positions on the perches. I would advise against perches unless you are keeping breeding stock to produce your own eggs.

The only other equipment needed are feeders and drinkers, and these are best suspended so that litter is not scratched into them. Below are photographs of a hanging drinker and a hanging feeder. Automatic watering is obviously the most convenient and this can be arranged to suit the particular housing. The first has the water laid on outside the house, with a hosepipe providing the flow. The birds have access through the wire, from inside. The second example shows completely automatic water systems with purpose-made drinkers.

There are some small farmers and smallholders who let their stock graze on pasture and where suitable, protected and clean land is available there is much to be said for this. Most people, however, find it more convenient to house their turkeys, away from the danger of foxes. Grazing is undoubtedly a cheap source of subsidiary feeding, but the grass declines from late summer onwards which is just the time that most people acquire their stock for the Christmas trade. It is essential to use land which has been free of all other poultry, otherwise the risk of disease transference is too great. It must be said that even if no domestic poultry has had access, wild birds can still introduce disease and parasites.

Acquiring Stock

The best time for the small producer to acquire stock is in mid-summer or early autumn, so that he can cater for the prime market – the Christmas trade. Six week old poults are readily available at this time, and are normally advertised in the poultry press. They are normally sold as A/H stock, meaning 'as hatched'.

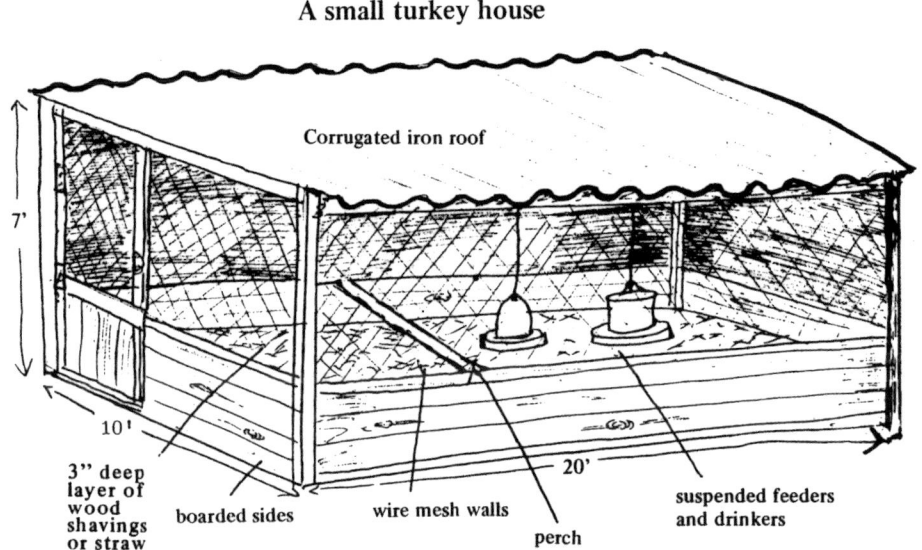

A small turkey house

This means that there will be both stags and hens (males and females) in the batch. Large producers usually separate the sexes at 6-7 weeks and house them in different quarters so that different rations can be given. This is because the stags grow at a faster rate than the hens and it is economically more appropriate to feed them different ratios. For the small producer this is not critical, and there is little to be gained from separating the sexes.

Stock acquired from specialist breeders will be healthy, and will have been injected against the disease Blackhead. In the early stages of life, turkey poults are liable to chills and other complaints, and the business of breeding and hatching is best left to the specialists. Once they are 6 weeks old, they are hardy, and have usually left any problems behind them.

The turkeys bred these days are a far cry from the wild birds which the Pilgrim Fathers encountered in America. They are heavier and produce far more meat than their ancestors ever did. While pure breeds of turkeys are still available, it should be emphasized that these are predominantly kept for show as fancy breeds. However commercial turkey breeders have begun to sell the traditional bronze feathered poults again and many poultry keepers will probably be interested to give these a try. The meat strains of selectively bred hybrids are 'maxi', 'midi' or 'mini' strains, depending on which sized oven they are catering for. They are white, not coloured as were the old breeds.

It is important to work out which strains you need before ordering. For example, if you have 'maxi' strains you may find that your turkeys grow too big and will be impossible to sell at Christmas. The size of the customer's oven that you are catering for is a vital factor, so do take the advice of your local supplier on this point.

Once they arrive, the poults should be transferred to their new quarters, with food and water available, and then left alone to settle in. Turkeys do have a tendency to panic and flap, and this settling in period is important. It is also important to enter their house in a quiet and unhurried way so that they do not panic and possibly damage their wings.

Many people have questioned whether turkey poults need to be de-beaked (have the tip of one mandible removed to stop feather pecking). My own view is that it is a repugnant practice and not necessary where birds have plenty of room and have enough of interest to keep the birds contented. The problem is that it is difficult to acquire poults which have not been treated in this way: most suppliers do this as a matter of course. However, if you feel strongly about it, try asking your supplier if he can arrange for you to have your birds which have not been de-beaked and see if he will oblige. The various organisations concerned with animal welfare may also be able to offer help on this matter. Their addresses are listed in the reference section.

Feeding

When the poults first arrive they may still be on a starter crumb ration which is the equivalent of chick crumbs. They can continue with this for about a week so that the change of diet does not coincide with the change to a new environment. After the starter ration it is normal to give a turkey rearers' ration, which is available from most livestock feed suppliers. This is a specially formulated feed to provide all the nutrients except water. It also contains antibiotics to prevent Blackhead and bacterial diseases, and a general growth promoter.

Most people are concerned at the level of additives in growers' rations and it must be said that, on a small scale, a more traditional pattern of feeding is more economical and sensible. Where turkeys are housed in a clean building, on a non-intensive scale and with plenty of ventilation, there is little likelihood of contracting disease. There is no need therefore to give them antibiotics in their feed. Especially as they will have been injected against Blackhead before they are sold by the supplier as poults. The small producer can therefore feed his turkeys on less intensive rations and thus cater for those customers who are interested to buy organically produced birds.

A good ration for growing turkeys is as follows:

> **1 part bran, 2 parts maize meal or oats,**
> **1 part fishmeal or soya extract and 2 parts wheat.**

This ration is fed two or three times a day, depending upon individual practice and routine. Crushed oyster shell grit must also be made available for the proper digestion of grain. Failure to do this will lead to problems with digestion, as well as a tendency for the birds to eat their litter.

Water should be available at all times.

It is important not to exceed the fishmeal ration, if given, otherwise the meat may receive a fishy taint. For the same reason, 'fishmeal' should be withheld for the last two weeks before slaughter. Many people now prefer to exclude it entirely and use soya extract or increase the maize meal content instead.

Photographs of various food elements that are mixed to make a wholesome feed for growth.

Turkey Crumbs

Wheat

Barley

Rearer Pellets

| Crushed Oats | Maize Meal |
| Bran | Crushed Oyster Shell |

Alternatively, it is possible to follow an even more traditional diet such as the one that was much used for Light Sussex table chickens in the past. My own method is to use barley meal which is widely available and relatively cheap. This is mixed to a coarse crumb consistency with surplus goat's milk. (Skimmed cows' milk is excellent, if available). This can be fed once, twice or three times a day, depending on your particular timetable, with grain given separately. The grain can either be a mixed grain or even wheat which is of course cheaper. Barley meal turns the skin and flesh a yellowish colour, anyone who remembers the farm poultry of pre-intensive days will also remember that table poultry was always this colour rather than the anaemic white it is these days. Comfrey leaves, cabbage heads and brussel sprouts or other vegetables can be hung in the house or run to provide interest as well as supplementary feed.

Killing and Plucking

The time to kill is basically the most convenient time for you. Leaving enough time for plucking, hanging, gutting, weighing and labelling, before meeting customers' orders for Christmas. It is usually a time for the whole family to pitch in and help with plucking, and it is surprising how, when the work is shared in this way, it is less of a chore. Listening to the radio and to music also helps.

Do not feed the birds for 12 hours before killing. But they should have access to water. They can be killed by neck dislocation, as is the practice with hens. And it is easier to do this with a helper: one to hold the bird and secure its wings, while the other does the killing. It goes without saying, that only those highly experienced should kill the birds. If the turkeys are to be sold as 'Norfolk dressed' birds, the head and feet should be left on after plucking. Where the birds are being sold gutted after plucking, the feet and head will be removed. In the latter case, decapitation can be achieved using a killing method, and there is a humane poultry killer available from suppliers. This is screwed into a permanent place on a wall, and operates similar to the guillotine principle.

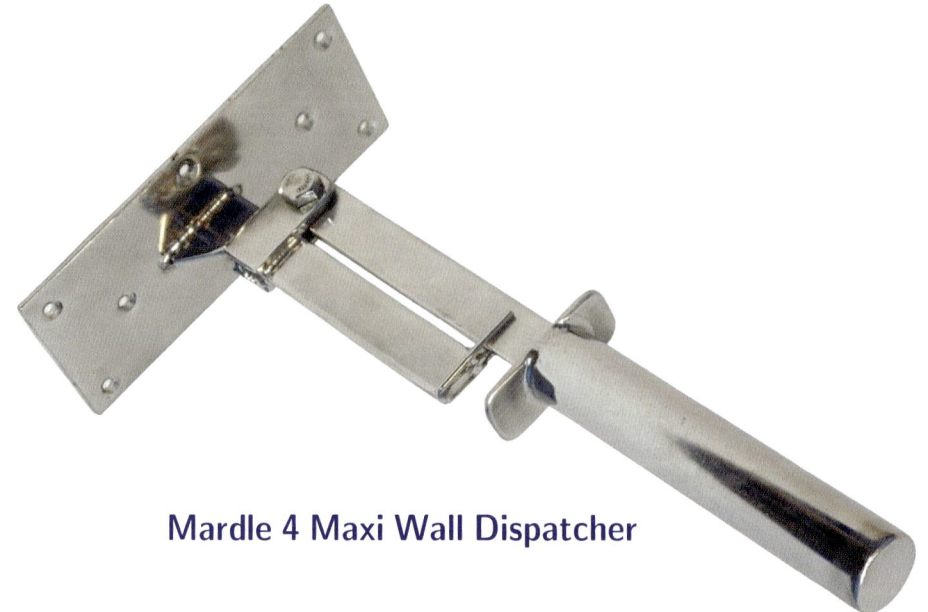

Mardle 4 Maxi Wall Dispatcher

There is no particular mystique about plucking, nor is there a 'right' way of doing it. The object is to remove as many of the feathers as possible, without tearing the skin. The way I do it is to pull out the primary wing feathers first – and for this a pair of pliers may be necessary – then, after finishing the wings, I go on to the legs. The body and down feathers come next, with particular attention being paid to any 'pin' feathers which are difficult to remove stubs. Here again, extra help may be required, this time with a pair of tweezers. Once plucking is completed, the birds should be left to hang in a cold outhouse where they are protected from the elements and dust. As most small producers produce turkeys for the Christmas market, there is usually no problem in finding a cold room in December.

Where a large number of turkeys are involved, it may be prudent to invest in a dry plucking machine. There are several available, costing some £500 upwards, and the manufacturers are normally pleased to arrange a demonstration for those interested in purchasing one. Mechanical flight pullers are also available with some starting at only £400; these remove the primary, secondary and tail flight feathers.

Dry Plucking machine from Bingham Appliances

Storey Poultry Supplies Twin Roller Flight Puller Machine

You may be able to find a demonstration of these machines on youtube.com

It is normal to weigh the birds before gutting, and this is achieved by using a good spring balance which can be suspended from a hook in the beam. The string around the bird's legs is hooked onto the hook of the spring balance and the weight recorded on a label which is then attached to the bird. Where birds are supplied to a local butcher, it is normal to leave the gutting to the butcher, who will do it for the individual customer.

Pluka P71

Solway Dry Plucker Machine   **Dry Plucking Machine**

Selling Christmas Turkeys

Ideally, you should have a clear idea of how many birds you expect to sell before you buy the poults. A certain amount of market research is necessary and butchers and retailers approached in good time.

You should aim to receive their orders in October. Where there is a wholefood butcher in your area, there is less likely to be competition, for demand at present exceeds supply. Most of the wholefood butchers are in the South of England.

Local markets normally have a sale of fresh turkeys just before Christmas, as well as auction sales arranged by the local auctioneers. These sales are normally well attended, and organically produced birds are particularly sought after.

For the small producer however, the biggest market is straight to the housewife, and it is up to the local producer to canvas for orders from the local populace. Where there is a local self-sufficiency group, it is a good idea to contact it so that information about the availability of local groups such as The Soil Association, The Henry Doubleday Research Association and The Friends of the Earth. Local Women's Institutes and Townswomens' Guilds are also worth approaching. All these organisations can be contacted by obtaining names and addresses from the local library. The local papers are useful advertising media, and it is a simple matter to approach them and ask them to do a small feature on your enterprise.

The British Turkey Federation, although primarily concerned with large producers are nevertheless interested in the small enterprise. The magazine 'Turkeys' is a useful source of information and help.

If you are selling turkeys direct at the 'farm gate'. You may need to have permission from the local authority for change of use from domestic to business use, if the enterprise is a large one and necessitates the use of a special building. Adequate parking facilities will be needed to cater for callers' cars, and an insurance cover is a good idea in case of accident.

The local DEFRA office will give advice, as well as the bank manager, accountant, local planning officer and insurance broker, not to mention a solicitor. Don't forget that if your house is mortgaged you will need to check with the mortgage company in case a change of use for commercial purposes invalidates your house insurance cover. There is usually no problem about this unless there happen to be restrictive covenants on the property. A good source of information for the small, part-time business is the Which Guide to Earning Money at Home, published by The Consumers' Association.

Suggestions for farm gate signage advertising turkeys for sale

For a small, part-time business, the question of VAT Registration will probably not arise, for the level of turnover is not likely to be achieved. It is best to talk to an accountant about this question, as well as income tax, for all income, whatever its source, must he declared. For this reason alone, it is in your interests to keep proper records including all costs such as – maintenance of buildings, purchase of stock, feedstuffs, electricity, equipment, supplies, petrol, telephone, stationery and all miscellaneous costs.

Profit Sheet

The Profit Sheet (Spreadsheet) shows an example of the profit potential of 100 poults purchased at six weeks old, it will be apparent that the costs are variable depending on the number of birds that are reared. The 100 poults is a price break from poult suppliers and 40 rearer sacks of feed is the ton rate, a price break from feed suppliers, to be delivered in one delivery. The supplier of poults for an order of 100 will normally supply 2 extra free, I have allowed losses of 5 birds during the growing on period. Please check that the poults have been vaccinated for Blackhead disease by the poult supplier.

Considerable savings can be made if some of the feed can be purchased as wheat or barley, direct from the farmer, 20+ sacks should be about £2.50 to £3.00 each. If chopped straw can be used as the flooring material, these from the local farmer will cost about £2.00 to £2.50 each, a good saving on the shavings.

If 100 poults are purchased, it would be advisable to run them together for a couple of weeks, then the stags (males) can be split off into a separate pen, the reason as explained elsewhere is that the stags grow faster and bigger than the female hens, it will stop bullying and ensure equal feeding opportunity for all the birds. If 40 poults or less are purchased, it is ok to leave all the birds together.

If the Birds are allowed to free range outside, they will eat as much green food and weeds as they want, plus any bugs they can find, this will supplement the feed bill. When the birds are brought back inside their pen permanently, if armfuls of weeds are collected and put in the turkey pens, the birds will eat these greens and plus the rearer feed will put on the weight ready for slaughter.

Another extra is to feed alfalfa hay in hay racks that the birds can reach inside their pen, off the ground to keep clean and edible.

The space requirements for 6 to 8 week poults should be 1.5 sq ft per poult, this means a pen of 10ft by 15ft will be ok for 100 poults.
From 8 to 12 weeks the floor space should be 2 sq ft per bird, from 12 to 16 weeks please allow 2.5 sq ft per birds and 16 weeks to full grown the space required is recommended to be 3 to 5 sq ft per turkey.

It will be noted from the Profit Sheet that more profit can be made if the full grown birds are sold from the farm gate or by selling direct to local customers. If the fact is advertised on the farm gate for potential passing customers to see that you can supply organic free range turkeys for Christmas, sales will be made, it is advisable that you take 50% deposits for any birds sold by this method.

Profit Sheet (Turkeys)

Cost Description	Qty	Each	Sub-Total
Shavings or Chopped Straw Bales	3	£9.50	£28.50
Poults Purchased (2 Extra Free)	100	£7.50	£750.00
Feed Crumbs Sacks	4	£10.00	£40.00
Feed Rearer Sacks	60	£7.50	£450.00
Selling cost if any	0	£0.00	£0.00
Delivery Petrol Litres	40	£1.20	£48.00
Total Costs			**£1,316.50**

Income Description **Wholesale and Butcher Sales**	Qty	lbs Per Bird	Total lbs	Sell Price Per lb	Sell Price
Turkeys Sold Small 'Mini'	21	12.00	252	£3.25	£819.00
Turkeys Sold Medium 'Midi"	18	17.00	306	£3.25	£994.50
Turkey Sold Large "Maxi'	16	23.00	368	£3.00	£1,104.00
Turkey Sold Extra Large "Maxi Plus'	7	27.00	189	£2.90	£548.10
Wholesale Income	**62**		**1115**		**£3,465.60**

Income Description **Farm Gate and Home Direct Sales**	Qty	lbs Per Bird	Total lbs	Sell Price Per lb	Sell Price
Turkeys Sold Small 'Mini'	15	12.00	180	£5.25	£945.00
Turkeys Sold Medium 'Midi"	10	17.00	170	£5.25	£892.50
Turkey Sold Large "Maxi'	7	23.00	161	£5.00	£805.00
Turkey Sold Extra Large "Maxi Plus'	3	27.00	81	£4.75	£384.75
Farm Gate Income	**35**		**592**		**£3,027.25**

Total Turkeys Sold	97				
Total lbs Weight			1707		
Total Income					£6,492.85
Total Costs					£1,316.50
Total Profit					£5,176.35

It must be noted that the above Profit Sheet is only a guide, if more turkeys are purchased to grow for sale, then there will be more income from sales. However the grower must have a ready market to sell the full grown turkeys.

Health

If stock is acquired from a reputable supplier it will be healthy and sturdy. And will have been injected against Blackhead. It depends upon the individual buyer whether it remains healthy, and emphasis must be placed on the availability of clean buildings, feeders, and drinkers. Litter should be dry, and if there is a build-up of damp manure, a clean layer of litter should be added. Every effort should be made to exclude rats from the house, for they are among the worst carriers of disease. The local DEFRA and local authority will both give advice on this question. If you operate from a commercially-registered farm or smallholding, the local authority will charge you for the services of their rat catcher, if he comes out to put down rat poison, but ordinary householders have this service free of charge.

Fowl plague and Newcastle Diseases are notifiable diseases; and any obviously serious condition which affects several birds should be brought to the attention of the vet immediately. If a notifiable disease is diagnosed, it must be reported to the local DEFRA office or the local police. Prevention is the best approach, and here again, the local DEFRA representative will give valuable advice.

The health of your birds must not be left unchecked, it is of great importance to monitor birds individually and thus ensure that you go on to produce healthy livestock ready for the table. To give you a better chance of identifying potential health problems and assist you to deal with any issues that may arise amongst your flock, we recommend a copy of Poultry and Waterfowl Problems by Michael Roberts. See our other titles available as listed in the back of this book.

Turkey Breeds

There are many varieties of Turkeys around the world, but for the purpose of farming them for human consumption we show you the most

British White

The white turkey has been recorded throughout the documented history of the turkey.

Mature Stag Weight: 12.7kgs (28lbs)
Mature Hen Weight: 9kgs (20lbs)

Norfolk Black

Farmers in the county of Norfolk showed great interested in keeping these birds for their fine meat and that is how the Norfolk Black turkey came to be named.

Mature Stag Weight: 11.35kgs (25lbs)
Mature Hen Weight: 6.8kgs (15lbs)

Bronze

Possibly the most popular and well-known turkey for the table and commonly sold in supermakets. It's the closest in colouration to the wild turkey.

Mature Stag Weight: 18.1kgs (40lbs)
Mature Hen Weight: 11.8kgs (26lbs)

commonly selected types for producing the best yield of meat and weight for maximum profit potential.

Narragansett

The Narragansett Turkey is a cross between a wild and domestic turkey.

Mature Stag Weight: 14.9kgs (33lbs)
Mature Hen Weight: 10.4kgs (23lbs)

Buff

There are now very few examples of the Buff to be seen. Bloodlines have been mixed with the Bourbon Red. Work is now needed to bring this variety of turkey up to standard.

Mature Stag Weight: 12.7kgs (28lbs)
Mature Hen Weight: 8.1kgs (18lbs)

Bourbon Red

The Bourbon Red is named after Bourbon County in the Bluegrass Region of Kentucky. It can be a very tame pet.

Mature Stag Weight: 14.9kgs (33lbs)
Mature Hen Weight: 8.1kgs (18lbs)

These varieties are now considered rare because other breeds are more commonly farmed due to meat yield and growth rates.

Pied / Cröllwitzer

This small turkey is not really selected for growth rate but can be great on smallholdings as an egg producer. Also a favourite bird for exhibiting.

Mature Stag Weight: 10kgs (22lbs)
Mature Hen Weight: 5.4kgs (12lbs)

Blue / Lavender

Developed from the Slate turkey, however the feathers are an even colour without any flecks. There is also a very pale Blue turkey called Lavender

Mature Stag Weight: 11.3kgs (25lbs)
Mature Hen Weight: 8kgs (18lbs)

Slate

Occasionally used as a commercial bird in niche markets, it is highly sought after as an exhibition bird. Now considered critically endangered rare worldwide.

Mature Stag Weight: 13.6kgs (30lbs)
Mature Hen Weight: 5.4kgs (12lbs)

Breeding

An increasing number of people are interested in breeding their own turkey poults, particularly where the older breeds such as Broad Breasted Bronze, Norfolk Black or Buff are preferred. Breeding stock or, in some cases, fertile eggs are available, the reference section will indicate such suppliers. During the hens six month production cycle up to 1500 poults can be sired by the stag. Turkey eggs take 28 days to hatch, either under the hen or in an incubator.

Breeding stock should be kept separate from other birds and the eggs removed from the hen when she starts laying. Keep them in a cool place such as an outhouse until you have enough to incubate. Although hen turkeys will go broody, like most female birds, they are not always reliable and most people prefer to use an incubator. Broody hens have also been used to good effect by some, but I prefer to incubate artificially in order to minimise the chances of disease transference.

Commercially, young ones are injected against Coccidiosis and if you prefer to do his, your local vet will advise on the procedure. However if the brooding and rearing conditions are kept clean, no damp litter allowed to build up and other birds kept clear of the area. There is a good chance that a small, non-intensively raised group of turkeys will not get it. If they are to be range reared then it is essential to take this precaution because wild birds can introduce disease. Artificial heat in a brooding system such as that used for chicks must be made available until the poults are hardy, for they are liable to die from chilling. Chick crumbs and fresh, clean water provide all their dietary needs for the first 6-8 weeks, when they

can gradually go over to an adult diet such as that described earlier. Where there is a surplus of turkey eggs, they can of course be eaten, but you will find that the shells are much tougher than those of hen's eggs. If you're entering for an egg-rolling contest over Easter, make sure you use a hard-boiled turkey egg – and you can't lose!

Number One Predator is the Fox

The fox is very cunning, but he is more cunning who catches the fox. (Pedro Calderon de la Barca)

The cunning of the fox is as murderous as the violence of the wolf. (Thomas Paine)

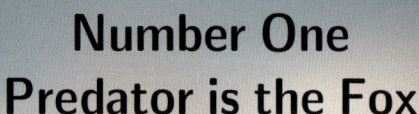

Vermin and Predators

This book would not be complete without writing about rats, mice and the fox. The rat is a clever resourceful rodent, they carry very infectious diseases for your young poults, rats if given the opportunity will eat vast quantities of your expensive feed. Every effort on your part must be used to keep them out of your turkey pen, by using ½ inch wire mesh, they cannot get through this small mesh. The edges of the pen should have ½ mesh buried up to 12" inches into the ground to stop rats and a fox from digging into your pen.

If rats are seen or droppings found, better to put down poison to kill them, trapping is also quite successful, personally I use both to keep any rats under control.

Live Catch Cage

**Fenn Trap MK4
Manufactured by DB Springs Ltd**

The Rat and his paw prints

Remember that prevention is better than cure, especially where financial investment or health are concerned. To help control these unwanted pests we reccommend a copy of Modern Vermin Control by Michael Roberts. See our other titles available as listed in the back of this book.

Reference Section

Publications
Starting with Turkeys by Katie Thear
Turkeys at Home by Michael Roberts
Incubation a Guide to Hatching and Rearing by Katie Thear
Poultry and Waterfowl Problems by Michael Roberts
Free Range Poultry by Katie Thear
Organic Poultry by Katie Thear
Farm and Smallholder Fencing by Michael Roberts
Gates and Styles by Michael Roberts
Incubation at Home by Michael Roberts
Country Smallholding magazine www.countrysmallholding.com
Smallholder magazine www.smallholder.co.uk
Practical Poultry magazine www.practicalpoultry.com
Fancy Fowl magazine www.fancyfowl.com
Home Farmer magazine www.homefarmer.co.uk

Organisations
The British Poultry Council www.britishpoultry.org.uk
DEFRA www.defra.gov.uk
Farm Animal Welfare Council www.fawc.org.uk
Humane Slaughter Association www.hsa.org.uk
National Farmers Union www.nfu.org.uk
Compassion in World Farming www.ciwf.org.uk
Poultry Club of Great Britain www.poultryclub.org
Rare Breeds Survival Trust www.rbst.org.uk
Utility Poultry Breeders Association www.utilitypoultry.co.uk
Rare Poultry Society www.rarepoultrysociety.co.uk
Anglian Turkey Association www.anglianturkey.org.uk
The Traditional Farm Fresh Turkey Association (TFTA) www.goldenturkeys.co.uk

Breeders
Kelly Turkey Farms poults 01245 223581 www.kellyturkeys.co.uk
Baron Turkeys 01928 716416 www.baronturkeys.co.uk
Rutland Organic Turkeys 01780 722009 www.rutlandorganics.co.uk
Irish Turkey Breeders www.irish fowl.com
Rookery Farm www.peeles-blackturkeys.co.uk

Source Turkey www.sourceturkey.com
Heritage Turkey Hatchery www.cacklehatchery.co.uk
Wonnacott Farm www.wonnacottfarm.co.uk
Cornwallturkeys www.cornwallturkeys.com
Birdtrader www.birdtrader.co.uk
Heritage Turkeys www.heritageturkeys.co.uk
Fenton Poultry www.fentonpoultry.co.uk
HockenHull Turkeys www.hockenhullturkeys.co.uk
Cyril-Bason www.cyril-bason.co.uk

Equipment
Mardle Products Ltd www.mardleproducts.co.uk Dispatchers
Green Valley Poultry Supplies www.Chickenhouse.co.uk
Quill Productions www.quillproductions.co.uk free range feeders
Solway Feeders Ltd www.solwayfeeders.co.uk Slaughter & plucking equipment
Parkland Products Ltd www.parklandproducts.co.uk Free range feeders
Rappa Fencing Ltd www.rappa.co.uk Electric fencing
Rutland Electric Fencing www.rutland-electricfencing.co.uk Electric fencing
Littleacre Products www.littleacre-direct.co.uk Housing
Potters Poultry www.potterspoultry.co.uk Housing
Hodgson Timber Buildings www.hodgsontimberbuildings.co.uk Housing
Smiths Sectional Buildings www.smithssectionalbuildings.co.uk Housing
Danro Ltd www.danroltd.co.uk Egg boxes and labels
Rooster Booster www.roosterbooster.co.uk 12 Volt Lighting Systems

Feeds
BOCM Pauls Ltd www.bocmpauls.co.uk
Hi-Peak Feeds www.hipeak.co.uk
Smallholder Feeds www.smallholderfeed.co.uk
W.H.Marriage & Sons Ltd www.marriagefeeds.co.uk

Services
Cliverton Insurance Brokers www.cliverton.co.uk Insurance
Swinton Insurance www.swinton.co.uk Insurance

STARTING WITH TURKEYS

The ideal introduction to keeping turkeys! This is a well illustrated and readable book covering all aspects of keeping, breeding, raising and showing turkeys on a small scale.

It describes the traditional breeds and how to cater for the free-range and organic Christmas markets. It is particularly useful for those who wish to produce free-range and organic turkeys for Christmas either for themselves or to an increasingly discriminating public at the farmgate.

The reference guide for those keeping turkeys

FARM AND SMALLHOLDER FENCING

A practical guide to permanent and electric livestock fencing on the farm and smallholding

This book has 18 chapters covering materials to use, constructing and maintaining fences and lots of useful tips.

Well constructed fencing will help to protect and keep livestock safe from predators. The right fencing should be selected for correctly controlling the livestock, allowing them to roam freely within their given area. Thought should also be given about access to allow you to easily service the animals needs. This might be housing, feeding, moving from one area to another etc.

Over 140 colour photos and illustrations.

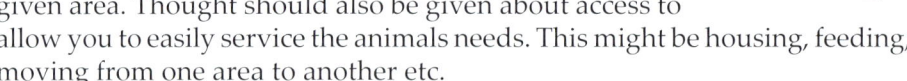

www.broadleyspublications.co.uk

Index

Acquiring Stock	15, 16	Introduction	7
Alfalfa Hay	26	Killing and Plucking	20, 21, 22
Bayle MP7 Dry Plucking	22	Mardle Dispatcher	20
Bingham Appliances	21	Narragansett Turkey	30
Bourbon Red Turkey	30	Norfolk Black Turkey	29
Blue/Lavender Turkey	31	Perches	12
Breeds	29, 30, 31	Pied/Crollwitzer Turkey	31
Breeding	32, 33	Pluka P71	22
British White Turkey	29	Poults	16
Bronze Turkey	29	Predators	34
Buff Turkey	30	Profit Sheet	25, 26, 27
Christmas Turkeys	23	Rat	35
Contents	2	Reference Section	36, 37
Defra	9, 24	Regulations	8, 10
Diseases	28	Selling Turkeys	23
Drinkers	13	Slate Turkey	31
Farmgate Signs	24	Solway Dry Plucker	22
Feeders	13	Storey Poultry Supplies	21
Feeding	17, 18, 19	Straw	11
Grazing	14	Traps	35
Health	28	Vermin	35
History	5	Wood Shavings	11
Housing	11, 12, 15		

Acknowledgments of images supplied by Shutterstock.com/

Page 1, Turkeys on Farm, Richard Mann
Page 3, Young Turkeys, Alaettin YILDIRIM
Page 4, Hand-drawn Turkey Illustration, mymaja8
Page 6, young turkey on a farm, blurry background, ene
Page 7, Flock of Turkeys, Richard Mann
Page 8, Farm Turkeys, Richard Mann
Page 9, A group of pasture raised turkeys, Tony Campbell
Page 10, Vector meat cuts thin line, robuart
Page 11, Wood sawdust texture, background, Guy's Art
Page 11, Straw texture, Roberto Sorin
Page 12, Cute turkey at the farm, AnastasiaNess
Page 13, Turkeys on farm (Yellow Feeder), Vipavlenkoff
Page 13, BEC 75 Automatic Drinker, BEC
Page 14, Turkeys, Richard Mann
Page 14, Young turkey on a farm, The Len
Page 14, Free-Range Bronze Turkeys, Ed Phillips
Page 18, Organic barley grains, keko64
Page 18, Grains of Wheat, Theerapong Silachan
Page 19, Oat Flakes, Viktor1
Page 19, Maize Meal, Dream79
Page 19, Miller's Bran, aga7ta
Page 19, Crushed Oyster Shells, Stockimo
Page 26, hay bale with blue tractor and baler, hookmedia
Page 26, square bales of fresh alfalfa hay, Jim Parkin
Page 26, Turkeys in a barn, Maria Dryfhout
Page 29, White turkey walking on the farm, Alexey Fyodorov
Page 29, Black turkey walking on the small farm, Dmitri Ma
Page 30, Narragansett Turkey, Nancy Kennedy
Page 31, A beautiful White and Black Turkey, mikeledray
Page 32, Domestic turkey, being reared on a farm, Erni
Page 32, Turkey and turkey-hens, ileana_bt
Page 33, Young turkeys on farm, Gordana Sermek
Page 33, Family of Turkey, Kemeo
Page 34, A fox turns and faces camera, Philip Ellard
Page 35, Rat in a metal catch trap, Suwatchai Pluemruetai

All images are the copyright and property of their respective owners. All rights reserved. No liability will be accepted for any error or ommission made within the credit or the reproduction of the images. No reproduction must be made without prior written consent, application for which should be made directly to the respective and actual copyright holder.

POULTRY AND WATERFOWL PROBLEMS

Provides the beginner and the experienced poultry keeper with a veterinary guide to problems, diseases and complaints of chickens, geese, ducks and turkeys. The book covers what may have caused the various conditions, what the condition could be and suggests ways to prevent or cure the problems.

Many of the problems are from the birds not being looked after properly, due to unhealthy living conditions, diseases which can be hereditary, or by purchasing diseased livestock from elsewhere, or the infection can be spread by wild birds, ducks or vermin.

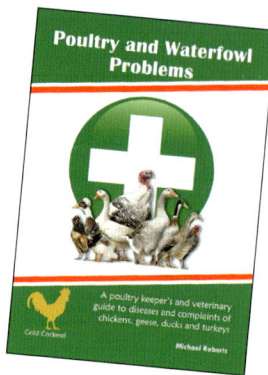

The book has been written from personal experience, if not by the author then from people who do have the relevant experience. Please note; poultry which have been kept in the correct conditions are normally healthy and robust. If experiencing problems with poultry the book will give a guide to nearly all conditions, but if a diagnosis is required please contact a local avian vet for advice.

POULTRY HOUSE CONSTRUCTION

This is a D.I.Y. guide to building hen houses, chicken coops and allied equipment such as show boxes, free range feeders and trap nest boxes.

It offers numerous simple designs for easy construction and the ongoing cleaning and maintenance where required. This book was written by Michael Roberts, an experienced poultry keeper who understands the needs of the birds and the desired practicality of the housing to allow efficency in servicing the needs of the poultry being kept.

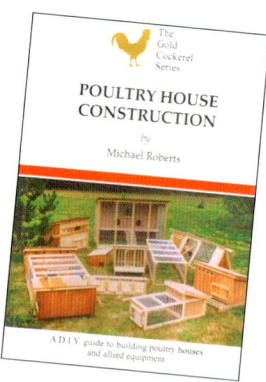

Items to build include: smaller chicken coop's to larger hen houses, runs, show boxes, free range feeders and trap nest boxes among others.

A great book allowing you to build your own affordable housing.

www.broadleyspublications.co.uk

MODERN VERMIN CONTROL

Vermin affecting poultry, game and the smallholder have always been there, today, we are more aware of the damage that they can cause to livestock and the diseases that they can carry. Anyone reading this book should use their best efforts to reduce and eradicate these pests that prey on livestock, or can bring mites, fleas and disease via food contamination.

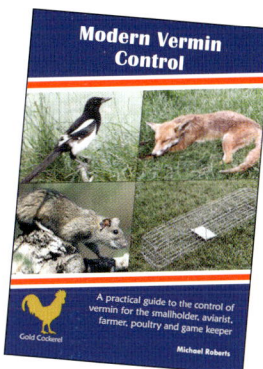

Many will agree that in keeping poultry one of the most dangerous predators to the birds will be the fox, a section on how to prevent fox entry, protection of hen houses and how to catch the fox is shown.

Covering all the most likely vermin, rats, mice, moles, squirrels, weasels, stoats, mink and carrion birds, the way to recognize their tracks, the various types of traps available and where to place, setting traps and baits to catch the different vermin species. Includes a section on the Law and which animals and birds are protected species.

INCUBATION - a guide to hatching and rearing

This jargon-free guide to natural and artificial incubation is designed for the small-scale poultry keeper and the commercial breeder.

Deals with the process of egg development and the conditions necessary for successful incubation, hatching and rearing.

To many breeders of poultry and other birds, the need to incubate eggs successfully is vital. Traditionally, broody hens were used - and this is still one of the best methods - but as they are not always available when needed, artificial incubation is the answer. Relevant to poultry keepers, waterfowl and game breeders, and anyone with an interest in domestic and exotic species of birds.

Species covered: chickens, ducks, geese, turkeys, guinea fowl, quail, pheasants, partridges, peafowl, ostriches, emus, rheas and parrots.

www.broadleyspublications.co.uk

STARTING WITH A SMALLHOLDING

A really useful book for anyone planning to buy a smallholding, or start a new smallholding activity.

Offers timely advice on how to choose the right property for different needs and circumstances, as well as sorting out priorities after moving in. It describes how activities make different demands on space, time, energy and money, with details of over 20 activities related to animals, poultry and the land. It also provides details of where further help and advice may be obtained.

This practical book provides essential guidelines to the care and welfare of livestock as well as to the efficient and enjoyable running of a smallholding.

ORGANIC POULTRY

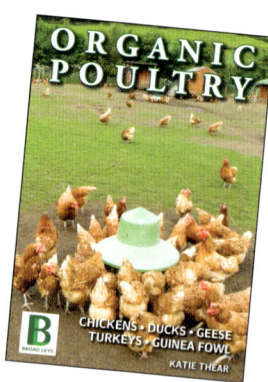

This unique book provides an excellent coverage of what is involved in managing organic poultry, including laying and table chickens, ducks, geese, turkeys and guinea fowl.

Includes research findings on the best way to cater for the innate needs of poultry and aims to clarify a situation where there is considerable confusion about organic standards.

A 'must' for those keeping poultry organically or considering converting to organic poultry.

List of Contents: Preface, What is organic?, Land conversion, Pasture, Housing, Feeding, Layers, Table chickens, Ducks, Geese, Turkeys, Guinea fowl, Health, Flock replacement, The organic enterprise, References, Index. This definitive work is by Katie Thear who was a well known author and poultry keeper. She advised on waterfowl standards for RSPCA Freedom Foods as well as for the Poultry Organic Standards.

www.broadleyspublications.co.uk

MAKING HOME MADE JAM, JELLY AND MARMALADE

A colourful and exciting book covering many classic jams, jellies and marmalades, but with a choice of many new recipes.

All recipes use recognised weights and measurements for UK, European, USA and Australasian countries.

The book covers hygiene, the equipment and utensils needed, recipes are in sections of Jam, Jelly and Marmalade. Each recipe is on a seperate page, together with the preparation of the fruit for cooking, the various stages are clearly set out to produce and create your chosen recipe. The book gives instructions on how to know when the various fruits are set and ready to pot, along with instruction on the potting procedure, using a funnel or spoon.

The many shapes and sizes of jars are shown, screw on lid colours, the use of cellophane with elastic bands to seal the jars and a section on the different types of cloth covers and why to use. Labelling is also covered. Includes a reference section of organisatons and suppliers from the UK, USA, Australia and New Zealand for the equipment and supplies needed to successfully make jam, jelly and marmalade.

MAKING HOME MADE CHUTNEY, PICKLE AND RELISH

Whether a beginner, or an experienced chutney maker who wants some new recipes, this book makes a great choice.

The book comprises many of the classic Chutneys, Pickles and Relishes, along with many new recipes. All use recognised weights and measurements for UK, European, USA and Australasian countries. Each recipe is on a seperate page, detailing the preparation of the vegetables for cooking and the various stages to produce and create your chosen recipe. It gives instructions on how to know when the various vegetables are ready to pot and the potting procedure. Also covered are hygiene, equipment and utensils, jar varieties, lids and available colours, cloth covers and labelling.

Make something different by adding a spice to your creation, the many spices are listed and the different tastes created.

Features a numbered index and very useful reference section of organisatons and suppliers from the UK, USA, Australia and New Zealand for the equipment and supplies needed to successfully make Chutney, Pickle and Relish.

www.broadleyspublications.co.uk

Titles available from Broad Leys
SPECIALISING IN POULTRY AND SMALLHOLDING BOOKS

An exciting collection of titles published by Broad Leys Publications. Thoughtfully written by knowledgeable hands-on experienced individuals, each with their own expertise and proven confidence in their field of interest.

TITLE	AUTHOR
Build Your Own Poultry House	Katie Thear
Cheesemaking and Dairying	Katie Thear
Incubation: A Guide to Hatching & Rearing	Katie Thear
Keeping Quail	Katie Thear
Making Home Made Chutney, Pickle and Relish	Nigel Bowerbank
Making Home Made Jam, Jelly and Marmalade	Nigel Bowerbank
Organic Poultry	Katie Thear
Starting with Bantams	David Scrivener
Starting with Bees	Peter Gordon
Starting with Chickens	Katie Thear
Starting with Ducks	Katie Thear
Starting with Geese	Katie Thear
Starting with Goats	Katie Thear
Starting with Muscovy Ducks	Nigel Bowerbank
Starting with Pigs	Andy Case
Starting with Sheep	Mary Castell
Starting with a Smallholding	David Hills
Starting with Turkeys	Katie Thear

The books listed above and any new titles are available from all good stockists or post-free by visiting www.broadleyspublications.co.uk

www.broadleyspublications.co.uk